Mirdes Fabiana Hengen
Rafael Alvise Alberti
Raquel Eloise Heck

Construction methods

Mirdes Fabiana Hengen
Rafael Alvise Alberti
Raquel Eloise Heck

Construction methods

From Brazil to New Zealand

ScienciaScripts

Imprint

Any brand names and product names mentioned in this book are subject to trademark, brand or patent protection and are trademarks or registered trademarks of their respective holders. The use of brand names, product names, common names, trade names, product descriptions etc. even without a particular marking in this work is in no way to be construed to mean that such names may be regarded as unrestricted in respect of trademark and brand protection legislation and could thus be used by anyone.

Cover image: www.ingimage.com

This book is a translation from the original published under ISBN 978-613-9-61833-0.

Publisher:
Sciencia Scripts
is a trademark of
Dodo Books Indian Ocean Ltd. and OmniScriptum S.R.L publishing group

120 High Road, East Finchley, London, N2 9ED, United Kingdom
Str. Armeneasca 28/1, office 1, Chisinau MD-2012, Republic of Moldova, Europe
Printed at: see last page
ISBN: 978-620-7-70592-4

SUMMARY

SUMMARY

Over the years, the civil construction industry has intensified, marked mainly by bringing one of the largest economic shares to the country, but the challenge is to achieve increasingly sustainable constructions and methods, a crucial point for the future of our society. The Brazilian people's culture and way of building emerged from the coffee economic cycle, which is still enjoyed today. The aim is to develop techniques capable of increasing productivity, optimizing time and generating less waste. Within these construction systems, the conventional masonry method and the method used in New Zealand, Wood Frame, are compared, pointing out all the advantages and disadvantages of each system, as well as checking their technical specifications and total costs. The results show that the public questioned is very accepting of innovative systems that fit their needs. In conclusion, the research presents results on a system capable of combining sustainability with its constructive benefits.

Keywords: Developed countries. Construction. Housing deficit. Sustainability. Innovation.

1 INTRODUCTION

For many years, according to the Brazilian Institute of Geography and Statistics (IBGE, 2016), the construction industry has been one of the largest economic sectors in the country, and is essential for the country's development. However, the Brazilian economy is currently in a downturn, and according to the IBGE, in the first quarter of 2015 it accounted for 8.3% of the Gross Domestic Product (GDP), falling to 6.2% in the first quarter of 2016.

The growing social imbalance together with the housing deficit has given organizations the need to modernize in the face of the country's development policy and assume a posture committed to socio-environmental responsibility. According to Reis, Fadigas and Carvalho (2005), the first major step was the Stockholm conference in 1972, which emphasized the importance of the environmental issue for the continuity of life and history.

Today in Brazil, the Ministry of Planning, Development and Management (PAC), developed by the Federal Government, seeks infrastructure investments to serve millions of Brazilians, being fundamental for the country's economy.

Adapting to this new economic reality, the comparison of an innovative Wood Frame method with the Conventional Masonry Building System guarantees new productivity tools and control methods, which according to Francklin Junior and Amaral (2008) along the way limit the existence of errors and waste that will have direct and immediate applicability in new constructions.

Based on its structure, the term Frame adopts dry construction as one of its main characteristics, which is made up of light, slender profiles spaced evenly along all the walls. It is this number of structural elements that makes the method able to withstand the acting forces, without the need to add materials with a high specific weight. This system, if well designed, has numerous advantages over the conventional Brazilian method, including speed and ease of construction, optimizing the use of materials since the process is industrialized, avoiding waste and increasing quality (CARDOSO, 2015).

The Lean Construction philosophy, together with the industrialized system, appeared to add value to the product, where the key point is to satisfy the customer through production efficiency, waste reduction, process clarity, cost reduction and, most importantly, the pursuit of perfection (ROLIM, 2012).

In view of the aspects observed, the question is to compare the construction methods and evaluate their performance, based on which will be the most efficient construction system: Conventional Masonry or Wood Frame?

2 BUILDING SYSTEMS AND THE PROCESS OF INDUSTRIALIZATION

In the last twenty years, the search for new construction methods and new technologies has intensified, marked mainly by materials and techniques with greater productivity and lower costs. These technologies found their place as conventional forms of production failed to meet the new demands of builders and consumers (MEDEIROS, 2012).

Innovative building systems are methods of construction that differ from conventional masonry. They dispense with bricks and frames, thus reducing the use of water on site. The main difference between innovative methods and conventional methods is that innovative methods, if well designed, manufactured and executed, can greatly reduce construction waste (BERTOLINI, 2013).

Some of these current and modern systems that are gradually gaining ground in Brazil are *drywall*, which is made up of plasterboard panels screwed onto steel profiles, and the *steel frame,* which has its structure in galvanized steel profiles (BERTOLINI, 2013).

An example of this is plasterboard walls, known as *drywall,* which are assembled, while conventional masonry walls are built from mortar and need time to dry, harden and acquire their proper strength (MEDEIROS, 2012).

Industrialized production, therefore, is based on the manufacture of materials and elements produced in industry, then adapted to the work. This process is associated with the principles of organization, planning, quality and control, aimed at minimizing waste, increasing productivity and reducing costs (DONIAK, 2014).

In civil construction, industrialization emerged to meet the needs of countries destroyed in the First World War, a process that streamlined activities, avoided greater waste and reduced the total cost of the work (MORAIS; LIMA, 2009).

2.1 Conventional Masonry

The use of masonry as the main building material has been with man for many years.

From 10,000 B.C., in Persian and Assyrian constructions, there are reports of the use of sun-dried bricks, and from 3,000 B.C., of bricks burnt in kilns (FRANCO, 1998).

Bricks are made by hand and are considered to be the oldest building material to date. Excavations in Jericho discovered the existence of bricks as far back as 6,000 BC, where they were eight to ten inches long. The permanence of this material in society was due to the ease of the raw material and the search for it (BROCK, 1994).

In the past, buildings were dimensioned without any structural concept, entirely by empirical sense, which is why we find walls of various thicknesses in old buildings, ranging from 0.30 m to 1.30 m (PEREIRA, 2005).

After a few years, according to Pereira (2005), in the early 1950s a thirteen-storey unreinforced masonry building was designed and dimensioned by the Swiss engineer Paul Haller, with 15 cm thick internal walls and 37.50 cm thick external walls, with a compressive strength of 30 MPa.

It was only with the Industrial Revolution that reinforced concrete and structural steel emerged, moving from the empirical method to the experimental method of design, revolutionizing the construction market and making this process more resistant (PEREIRA, 2005).

According to Bastos (2006), reinforced concrete emerged in the mid-18th century to meet the needs of concrete strength and steel durability against corrosive agents. After the 19th century, it was called reinforced concrete.

It was in France that the first boat made of reinforced concrete was developed, by the Frenchman Lambot, from thin iron mesh completed with mortar (BASTOS, 2006).

2.1.1 Brazilian History

The use of bricks in Brazil became popular during the coffee economic cycle, starting with works directly linked to the agricultural products of the time (VILLAR, 2005).

The way we build today is still deeply related to the culture of the people who migrated here and brought their construction methods and practices with them. This inheritance from the Portuguese, Italians and Spanish has turned Brazilians into major consumers of masonry (bricks and blocks), mortar coverings, ceramics and stone (MEDEIROS, 2012).

At the beginning of the 19th century, reinforced concrete houses and townhouses began to appear in Brazil. After a decade, the Frenchman François Hennebique planned and calculated a nine-meter span bridge, which was built in Rio de Janeiro by the constructor Echeverria (BASTOS, 2006).

In Brazil, the "conventional method of construction" is the construction method most commonly used in today's buildings. Reinforced concrete structures and ceramic brick masonry represent the most common construction system in Brazilian buildings (ARAÙJO, 2012).

2.1.2 Reinforced concrete

The Conventional Masonry construction system is characterized by its reinforced concrete structure that supports it, foundation, pillars, beams and slabs, while the masonry only appears to seal the construction (SANTOS, 2014).

According to Correa (2013), the structural functioning of the conventional system starts with the "slabs", which are supported by the "beams", which are supported by the "pillars" and finally their total weight is unloaded "on the foundations", which are then evenly relieved by the soil.

Porto and Fernandes (2015) state that slabs are structural elements that transmit loads to the beams, which carry the loads to the pillars, which in turn propagate them to the foundation.

Concrete is made up of large aggregates (crushed stone, pebbles), small aggregates (sand), binders (cement), water, mineral additions and additives (accelerators, retarders, fibers, dyes), a material widely used in construction (PORTO; FERNANDES, 2015).

According to Schakelford (2008), concrete is a combination of fine aggregates (sand) and coarse aggregates (gravel), which are natural materials such as wood. The combination of the two is chosen to efficiently fill empty spaces. The matrix surrounding the aggregates is cement.

The aim of reinforced concrete is to meet the needs of concrete and steel, since one is responsible for the durability of the other. As concrete is a material with low tensile strength, the addition of steel to resist high tensile stresses is essential. On the other hand, steel is a material that suffers from corrosion and, with the application of concrete, the material is fully protected, so that both resist the stresses required in solidarity (BASTOS, 2006).

2.1.3 Foundations

According to NBR 6122 (ABNT, 1996, p. 02), a shallow foundation is "the foundation element in which the load is transmitted to the ground by the stresses distributed under the base of the foundation, and the depth of settlement in relation to the ground adjacent to the foundation is less than twice the smallest dimension of the foundation".

The traditional element for shallow foundations, also known as shallow or direct foundations, are footings, which transmit vertical loads and other actions to the ground. The footing is known as a reinforced concrete structural element, designed to withstand the resulting tensile stresses (BASTOS, 2016).

The deep foundation defined by NBR 6122 (ABNT, 1996, p.02) is a "foundation element that transmits the load to the ground either by the base or by its lateral surface or by a combination of the two, and its tip or base must be set at a depth greater than twice its smallest dimension in plan, and at least 3.0 m".

According to Barros (2011), deep foundations can be piles and pipes, which transmit the load through lateral friction between the element and the ground.

2.1.4 Walls

Masonry walls are made of ceramic bricks with the addition of mortar, and are designed to fill the gaps in reinforced concrete, steel or other structures. In this way, the walls must support their own weight, the weight of variable loads and accidental loads (THOMAZ et al., 2009).

According to Oliveira (2012), the closure consists of the walls themselves and is made of perforated ceramic bricks for sealing, laid in half steps, and it is extremely important to have the rows level, straight and aligned.

Mortars are used which consist of sand, cement, which has the characteristic of adherence, mechanical resistance and watertightness in the joints, and hydrated lime, which has the function of retaining the water, ensuring a lower modulus of deformation in the wall structure (THOMAZ et al., 2009).

When framing doors and windows, lintels and counter-lintels must be designed to avoid cracks caused by the loading of the openings (OLIVEIRA, 2012).

The sound insulation of hollow ceramic blocks is 42 decibels. There are several models of sealing blocks, the most commonly used are those with six, eight or nine equal holes, with the nine being the most recommended, allowing the opening of slots to embed the pipe with only one line reached, and the other two are left intact, facilitating better wall stability (YAZIGI, 2013).

2.1.5 Coating

Various types of cladding are used in masonry constructions, including plaster, stone, paint and ceramics, among others. What is taken into account is where the cladding will be applied, whether it will come into contact with humidity, whether it will be in wet or dry areas or whether it will be exposed to the outside environment (OLIVEIRA, 2012).

It is necessary to remove dirt, oil, nails and other materials left over from the bricklaying stage in order to start preparing the base. All gas, water and sewage pipes are also tested under the recommended pressure before the cladding begins (YAZIGI, 2013).

After the masonry has been laid, it is important to plaster the walls, plaster to ensure a better finish and, in some cases, plaster. This stage is followed by the internal and external cladding process. In some areas, a layer of waterproofing is applied to ensure that no water can get through (OLIVEIRA, 2012).

2.1.6 Electrical and hydraulic installations

There are countless resources available for the construction of water, sanitary and electrical systems. The most common is the use of *shafts*[1] . Normally the pipes are embedded in the masonry itself (in the case of blocks with vertical holes) and the horizontal pipes, whenever possible, pass through the slabs (THOMAZ et al., 2009).

In most cases tears are made in conventional walls, taking a considerable amount of time to open and close, increasing labor and material waste (OLIVEIRA, 2012).

2.1.7 Coverage

To make the roof, a wooden structure is usually used to receive and support the weight of the roof tiles. This structure is basically made up of scissors, purlins, rafters and slats (OLIVEIRA, 2012).

According to Yazigi (2013), a timber structure is made up of shears or purlins and main beams. This structure consists of a main frame and a secondary *frame.*

The roof can be finished with a variety of materials, including ceramic tiles, galvanized steel, aluminized wood, concrete tiles, corrugated fibre cement sheets, PVC and *fiberglass*[2] . These must resist the action of water, wind and other inclement weather (MOLITERNO, 2010).

2.2 Wood Frame

According to Pfeil and Pfeil (2003) wood is considered to be the oldest building material used by man, as it is relatively common in nature and has one of the main characteristics for transportation and handling: lightness. The main engineering works were built from stone and wood.

Although it was a long time coming, it wasn't until the 20th century that techniques applied to wooden structures were stipulated. After the Second World War, the subject gained new projects in a wide variety of structural forms (PFEIL; PFEIL,

---

1 *Shafts* are technical closures installed to enclose pipes, facilitating later adjustments (SWELL, 2013).
2 *Fiberglass roof* tiles are made from fiberglass and can be translucent, milky or colored (HIDROFIBER, n.d.).

2003).

In terms of functional performance, buildings have fundamental requirements in terms of structural strength and lightness, sealing, durability, thermal-acoustic insulation, visual and economic aspects. Because it meets these requirements, wood also stands out for being a renewable, abundant, highly sustainable source of energy that costs less to produce (SILVA, 2004).

With the Industrial Revolution, the market opened its doors to innovations in machinery, accompanied by population expansion, where the *Wood Frame* construction system gained a foothold. The prefabrication of strips of reforested and treated wood created their own forms, from wall panels to roofs combined and/or coated with other materials (SOUZA, 2013).

Wooden construction techniques have been known since prehistoric times and have been used in various periods and places around the world, especially in North America and northern Europe. The first prefabricated wooden houses appeared during the colonial period, in 1578, transported from England to Canada, and a few years later, in 1624, built by Edward Winslow, taken from England to Massachusetts (SOUZA, 2013).

Wood Frame has become well known in developed countries such as the United States and Canada, due to its ease of execution, its high degree of industrialization and the reduction of deadlines and labor costs (SACCO; STAMATO, 2008).

According to Bertolini (2013), *Wood Frame* revolutionized the American construction market, especially housing, where its great advantage was its rapid execution and cost reduction.

Plywood made from the *Pinus taeda* or *Pinus Elliottii* forest species for *Wood Frame* construction came from the southern region of the USA and began to be used in Brazil after the 1970s and 1980s (ABIMCI, 2013).

2.2.1 Brazilian History

Wood framing arrived in Brazil with the German immigrants, who brought it with them and determined it as a half-timbered structure (where its structure was basically made of wood with its pieces fitted together horizontally, vertically and inclined). This process had its greatest potential in the state of Santa Catarina, especially in the Blumenau region (FUTURENG, 2012).

On August 24, 2016, Brazilian civil construction took a big step forward in the new construction method, making history. This date marked the first three-storey building, built in 64 hours with high-tech sustainable Wood Frame, in Araucària in the state of Paranà (TECVERDE, 2016).

The most commonly used solution in the *Wood Frame* construction system are lightweight structures, with the load received from the entire building and distributed evenly along the ground, known as a *Radier as shown in* Figure 1. This has the characteristics of being a continuous, solid concrete slab, and is therefore a shallow or direct foundation. On the upper floors, *Oriented Strand Board* (OSB) is used as a subfloor and to ensure sound insulation, carpet or floor coverings are used with an intermediate blanket over the panels (SANTOS, 2012).

Figure 1 - Radier foundation

Source: Silva et al. (2012).

The structure is characterized by having load-bearing walls responsible for supporting the second floor. Once the first panel is stable, new load-bearing walls are erected over the floor, and so gradually up to the roof, and it is capable of completing works of up to four floors (FERREIRA, 2014).

The panels are made up of vertical wooden uprights, spaced 40 cm or 60 cm apart as shown in Figure 2, making it possible to use drywall and OSB together. The panel is closed with two wooden guides, one lower and one upper. Once all the panels on the second floor have been assembled on the foundation platform, a second guide is nailed over the upper guide, making all the partitions on the first floor solid. Galvanized nails are used for this connection, as they should have a long life (FERREIRA, 2014).

Figure 2 - Wood Frame composition

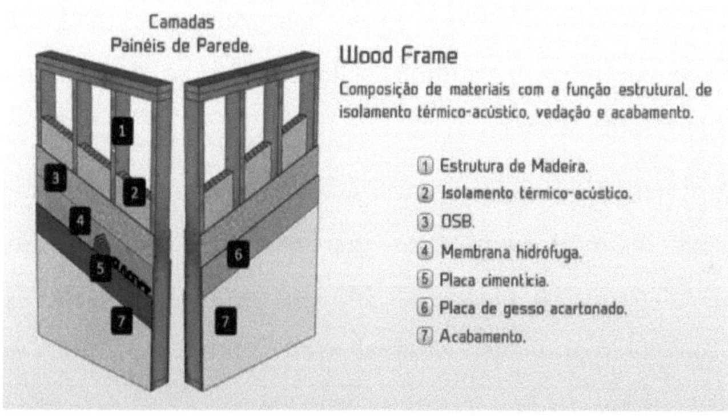

Source: Tecverde Network (2016).

For frames, doors and windows, the uprights in the panels should never be removed, but moved to provide greater lateral support, and one more should be added at the top to provide support for the use of lintels. Two more pieces of upright should be placed at the bottom, 38 mm less than the bottom height of the opening, so that it receives another piece of horizontal upright. Remember that these must be spaced the same as the load-bearing walls for OSB and drywall (FERREIRA, 2014).

OSB sheets are made from reforested wood, which is oriented in three crossed layers, perpendicular to each other, bonded with resins and pressed. The thickness of the OSB board is determined by the spacing between uprights and the type of cladding. Typically, 11.1 mm panels are used for walls and roofs and 18.3 mm panels for floors and slabs (SILVA, 2010).

2.2.4 Coating

The thickness of the OSB sheets for the cladding is determined according to the project, respecting the spacing of the uprights and the type of cladding to be used. A water-repellent membrane is applied to the OSB panels to protect the walls from external humidity, allowing water vapor to escape from inside the house (SILVA, 2010).

The function of external cladding is somewhat aesthetic, but especially against the action of the sun. External walls can be clad using different systems, from wood, steel or PVC *siding*, developed especially for the *Wood Frame* construction method, but reinforced mortar, exposed brick or even cementitious slabs can also be used, giving a finish similar to masonry (TÉCHNE, 2008).

For internal cladding, as illustrated in Figure 3, *drywall* guarantees optimum acoustic performance, and if mineral wool is applied to the inside of the wall, its thermal/acoustic performance gradually increases. Mineral wool or insulating blankets can be installed on internal walls, external walls and roofs (SILVA, 2010)(TÉCHNE, 2008).

Figure 3 - Composition of structure and cladding: a) Interior of wood frame building; b) Wood frame clad in drywall

Fonte: Net Zero Energy Cape Cod (2011).

2.2.5 *Electrical and hydraulic installations*

As *Wood Frame* is an industrialized construction method, the walls, as soon as they leave the factory, already have the electrical and hydraulic installations in place, shown in Figure 4, ready to be joined to any type of finish (ZAPARTE, 2014).

Figure 4 - Electrical and hydraulic installation still in the factory

Source: Tecverde Network (2016).

The water installations are installed between the uprights of the load-bearing walls and between the ceiling and the floor joists. The installations are designed and executed with PVC or PEX pipes[3] , the sewage pipes, which in most cases have a diameter of 100 mm, make it difficult to embed the sewage in the walls, and the use of *shafts* is essential (ZAPARTE, 2014).

The industry offers electrical and hydraulic materials produced especially for *drywall* and *framing*, such as electrical boxes and flexible systems known as PEX pipes, made of cross-linked polyethylene, which have the characteristics of resisting high temperatures and can be used for both cold and hot water (SILVA, 2010).

The electrical installations are carried out through conduits built into the wall and over the ceiling, conduits that are factory-fixed. The power wires are routed through the conduits after the walls have been fixed in place (ZAPARTE, 2014).

2.2.6 Madeira

Wood is an organic and natural material with a cellular structure and renewable raw

---

3 PEX pipes are flexible cross-linked polyethylene pipes, making it easier to make bends and reducing connections (NAKAMURA, 2013).

materials. It has characteristics and physical/mechanical properties that make it a material capable of providing resistant, safe, durable, comfortable and versatile buildings (CAMPOS, 2006).

Because it is a natural material, wood has considerable flaws that can affect its mechanical strength, such as knots and cracks, but on the other hand, these unfavorable aspects are easily solved with the use of industrialized products, the raw material is properly treated, resulting in suitable structures (PFEIL; PFEIL, 2003).

Knots come from branches on tree trunks and therefore reduce the strength of the wood by interrupting the continuity and direction of the fibers. In general, knots act more in tension than in compression (CALIL JUNIOR; LAHR; DIAS, 2003).

When compared to traditional building materials such as masonry, steel, among others, wood has a low weight and low energy consumption, as well as high mechanical resistance (CAMPOS, 2006).

One of the physical properties of wood that is relevant to projects is its moisture content. When wood is dried, the water contained in the hollow cells evaporates, reaching the point where the fibers are saturated. At this stage, the water inside has evaporated, but the cell walls are still saturated. This point is equivalent to 30% of its moisture content, so the wood is said to be semi-dry. After drying, the wood comes into equilibrium with the air and is then called air-dried. In Brazil and the United States, 12% is used as the standard reference unit (PFEIL; PFEIL, 2003).

Wood undergoes shrinkage or swelling, properties which are directly linked to its degree of humidity, so the drying process is extremely important, since over time the wood may show defects such as curling, bending and twisting (CAMPOS, 2006).

According to Calil Junior, Lahr and Dias (2003), the anatomical characteristics of wood require that the drying and storage process be carefully conducted, because if there are any defects in the pieces, they will restrict or even prevent their structural use.

Wooden parts are susceptible to biological attacks and the action of fire, but chemical treatments and fire retardants can relatively increase their resistance. Because wood is characterized as a combustible material, it is often considered to have low resistance to fire. However, when treated properly, it performs very well (PFEIL; PFEIL, 2003).

According to Calil Junior, Lahr and Dias (2003), although wood has a high flammability rate, structural parts have a better performance when exposed to high temperatures than other materials. The surface carbonation of the pieces becomes a kind of "thermal insulation barrier", so it is a poor conductor of heat, which does not compromise the central region of the pieces, unlike steel, which would have already collapsed (flowed), which even though it is not flammable is an excellent conductor of heat.

Another fundamental characteristic of wood in relation to the action of fire is the fact that it does not distort when exposed to high temperatures, making it difficult to ruin the structure, which despite being corbonized still has bearing capacity (CALIL JUNIOR; LAHR; DIAS, 2003).

Brazil is a country with a strong forestry vocation. Brazil's forest cover corresponds to 519.3 million hectares, of which 98.6% (512.1 million hectares) is made up of native forest. Planted forests account for only 1.4% (7.2 million hectares) of Brazil's forest cover (ABIMCI, 2013).

Of this total of planted forests, 71% (5.1 million ha) are eucalyptus, 22% (1.6 million ha) correspond to *Pinus,* used in *Wood Frame* constructions, and the remaining 7% (521 thousand ha) refer to other forest species (ABIMCI, 2013).

Pinus is practically concentrated in the southern region of Brazil, where it covers 85% (1.32 million ha). This species grows very well in places with a mild winter climate, with constant humidity throughout the year, especially in the states of Paraná and Santa Catarina, where the largest number of companies working with solid wood products are concentrated (ABIMCI, 2013).

2.2.7 Coverage

The *Wood Frame* system applied to roofs begins with the manufacture of wooden trusses or roof beams with a low inclination, followed by the installation of the base (OSB panels or plywood sheets) and the application of the moisture barrier, ending with the roof (ZAPARTE, 2014).

Shingle roof tiles[4] are the most widely used roof tiles in the *Wood Frame* system, as they are light and flexible. To use these tiles, the roof needs to be braced with OSB sheets. In warmer regions, the system is more efficient as it facilitates adhesion (ZAPARTE, 2014).

2.3 Advantages and Disadvantages of Building Systems

As for comparing the production systems, Wood Frame and Conventional Masonry, Table 1 provides some information.

Table 1 - Summary of comparisons between building systems

Features	Wood Frame	Alven. Conv.	Comments
Lightness	Z	X	*Wood frame* constructions together with OSB sheets guarantee lightweight structures compared to concrete (REVISTA DA MADEIRA, 2013). The main disadvantage of reinforced concrete is its own weight ($2.5t.m^3$) (PORTO; FERNANDES, 2015).
Agility	Z	X	In terms of consumptive agility, a 200m house[2] made of conventional masonry could take 12 months to complete, while the same house built with *Wood Frame* takes only three months to build (REVISTA IPÊ AMARELO, 2016).
Productivity	Z	X	In the conventional system, there is no systematic sequence of execution, and work is often interrupted on site. *Wood Frame,* on the other hand, is an industrialized system adapted to the Lean Construction philosophy, "less everything", fewer failures, less manpower, less waste (FARAH, 1996

---

4 *Shingle roof* tiles are made of asphalt, covered in volcanic rock, and are highly durable and aesthetically pleasing (DOCE OBRA, n.d.).

			apud VIVIAN; PALIARI; NOVAES, 2015).
Electrical and hydraulic installations	Z	X	The electrical and hydraulic building systems can be the same as the masonry systems, but what changes is the construction process: in the conventional system there is more labor, more waste and less practicality compared to the industrialized *Wood Frame* system *(CALIL; MOLINA, 2010)*.

Features	Wood Frame	Alven. Conv.	Comments
Coverage	√	X	Prefabricated shears have greater advantages than those produced on *site*. *They* are made to withstand snow loads, carefully designed and controlled, reducing the amount of waste and increasing its quality respectively (ZAPARTE, 2014).
Energy consumption	√	X	Masonry construction generates higher energy consumption and polluting gases during the mining of raw materials. *Wood Frame,* on the other hand, reduces CO^2 (carbon dioxide) emissions into the atmosphere by up to 73% (THEODOZIO JÛNIOR, 2006).
Structural capacity	X	√	The disadvantage of *Wood Frame* compared to conventional masonry is that it is only capable of completing projects up to four floors (FERREIRA, 2014).
Waste generation	√	X	As *Wood Frame* is a system that enables dry construction, it generates less waste compared to conventional constructions (MARTINS; ECKER, 2014).
Thermal and acoustic comfort	√	X	The thermal and acoustic comfort of *Wood Frame* is superior due to the use of suitable materials that guarantee good insulation (CALIL JUNIOR; MOLINA, 2010).

Source: Own authorship

3 LEAN CONSTRUCTION

The ideas of the new production philosophy emerged in Japan in the 1950s, applied to the Toyota production system, with the aim of eliminating stocks, waste and other techniques. The company's quality has evolved extensively, spreading the technique to American and European companies. A statistical study carried out in the USA and Europe ensured that the philosophy was characterized by "less everything", less human effort, less time, less waste (KOSKELA, 1992).

Nowadays, organizations are faced with high rates of technological innovation and a high level of competitiveness, each with its own assumptions to ensure an attractive look in the job market, thus addressing people management as one of their main focuses, which aims at the concept of quality of their products and the production process as a collective benefit, thus improving their performance as a whole (FERNANDES, 2015).

Construction systems that have a lower environmental impact, are more economical and quicker to build, reduce waste and consume less energy are the keys to the sustainable development that the country's future awaits. This is the path that the *Wood Frame* construction method follows, forming part of the Sustainable Energy Construction (CES) system, complying with construction standards that are essential for its realization (LEITE; LAHR, 2016).

Building sustainably is a pressing need. The use of resources during construction and throughout its use has never been so important. However, there is still a considerable way to go to meet the needs and practices related to the subject (MEIDEIROS, 2012).

In the construction industry, a preconception has been created that budget overruns and imperfections in construction mechanisms and materials are part of reality, which in turn can only be minimized with technological advances in the field. This situation leads to delays, cost variability and planning errors during construction work (PINTO, 2012).

The *Lean* philosophy proves that it is possible to achieve favorable results in a short space of time. Reducing waste and wastage, controlling and managing the activities required are all tools that can help to ensure that all the company's activities are carried out correctly (PINTO, 2012).

4 discussions

4.1 Acceptability questionnaire:

A questionnaire, Appendix A, assessed the acceptability of the new *Wood Frame* construction method, comparing it with the conventional masonry construction method.

The questionnaire was quantitative, with clear and objective questions, and the information was collected from a sample of 221 people. The study was carried out via the Google Forms platform and disseminated on social networks in order to reach a wider audience.

Among this sample of people, 79 were male and 142 were female, i.e. 64% of the respondents were female.

Following on from the questionnaire, analyzing the age group, the majority of respondents were between 19 and 25 years old, amounting to 43% of the 221 analyzed. Figure 5 shows that young people aged between 19 and 25 and between 26 and 35 formed the largest part of the graph, because they are more connected to social networks, among other things. The over-50s, on the other hand, accounted for only 8% of the responses, which is due to the fact that this group of people is often not connected to the digital environment.

Figure 5 - Age of respondents

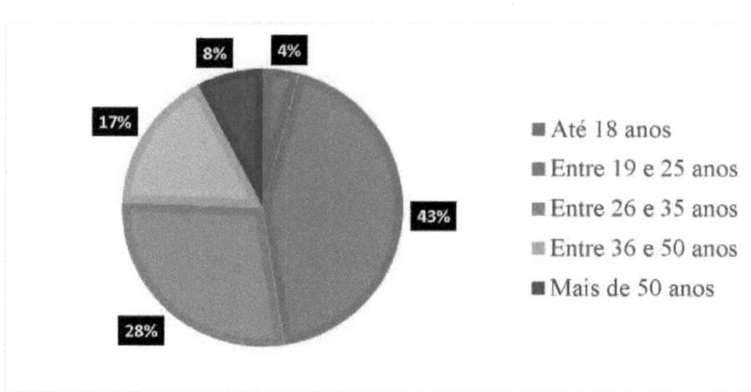

In order to reach out to the regional population to assess the acceptability of building with this different method. As can be seen in Figure 6, 41% of the respondents, or 91 people, are from the municipality of Sao Joao do Oeste/SC, and 64 respondents are from neighboring municipalities. A further 30% came from places such as New Zealand, Florianópolis and Chapecó. These respondents were chosen because they were already familiar with the *Wood Frame* construction system.

Figure 6 - City of respondents

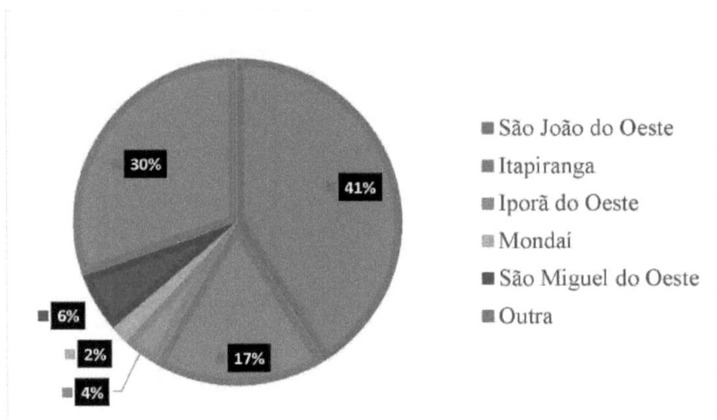

In this way, the questionnaire was sent to people from New Zealand and those who had already done exchanges in other countries, such as the United States, Canada and Germany, because they had lived in *Wood Frame* houses. Of the 221 respondents, 26 had lived or stayed temporarily in *wood frame houses*.

In order for the survey to be coherent and to find out how people really feel about the system, it was necessary to assess whether the population knew about the *Wood Frame* construction method, and for this, the questionnaire included a video explaining the system, according to Regli (2012).

Given this explanation, it can be seen that a significant proportion of the sample did not know about *Wood Frame* houses, totaling 70% of the sample surveyed. As mentioned by Villar (2005), one of the reasons for this is the culture introduced by the coffee economic cycle, where bricks became more popular than other building materials.

Innovative building systems are methods of construction that differ from conventional masonry. They dispense with bricks and frames, thus reducing the use of water on site. The main difference between innovative methods and conventional methods is that innovative methods, if well designed, manufactured and executed, can greatly reduce construction waste (BERTOLINI, 2013).

Based on this quote, the question was raised about the methodology of dry construction, that is, in this case, *Wood Frame,* where the only stage of the work that uses water is the *Radier* type foundation, the rest of the work being carried out with materials that do not require the use of water, such as: wooden panels, plasterboard sheets (*drywall*), glass wool and waterproof membrane.

Sustainability related to construction as attitudes to saving resources and making better use of what is consumed, water and energy for example, are used in the *Wood Frame* system where only water is used in the foundation, and the energy used to produce the wood is much less than that of other materials.

This also shows that the system is not very widespread, since the largest proportion of responses was negative, with 190 people having no knowledge of the stages of the construction system, nor of the materials used throughout the project (figure 7).

Figura 7 - Sustainability in building systems

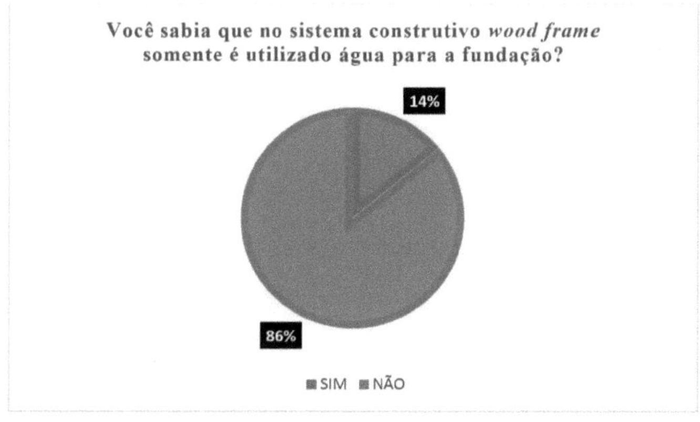

According to Bertolini (2013), *Wood Frame* revolutionized the American construction market, especially housing, where its great advantage was its rapid

25

execution and cost reduction.

Ecker and Martins (2014) built a housing complex of 339 houses using the *Wood Frame*, Steel Frame and Conventional Masonry systems, in order to compare the costs of one system and the other. Using budgets, they concluded that the direct costs of each 50m *Wood Frame* house[2] is R$39,786.66, of which R$9,635.40 is labor, and the direct costs of each Conventional Masonry house is R$48,465.39, totaling R$19,462.04 in labor. It can be concluded that the material prices between the two systems are practically equivalent, the final difference being the labor.

The survey confirmed that one of the main concerns when carrying out a project is the cost difference between one system and another. Among the respondents, 61% believe that the final cost of the project would define the choice between one system and another (Figure 8).

Figura 8 - Total construction costs

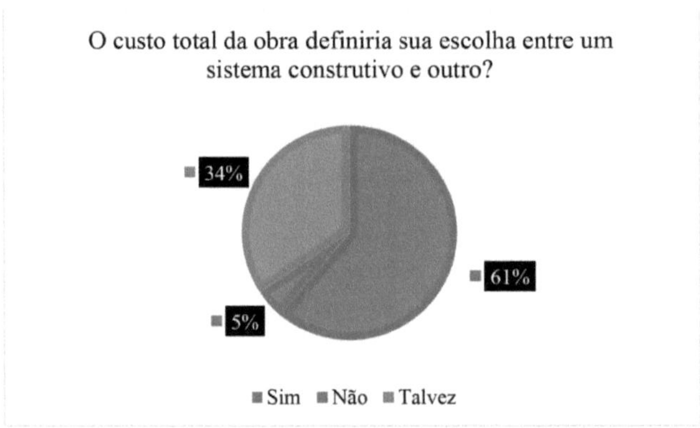

To accompany the reasoning behind the total cost of construction work and its importance, the final total time it takes to execute the project properly must also be taken into account. The vast majority of respondents, 94%, believe that speed of execution is an item to be taken into account when closing the deal (Figure 9).

Ecker and Martins (2014) created a timeline to compare the time taken to complete each construction system. They concluded that 339 *Wood Frame* homes would be finished in 14 months, while completing 339 Conventional Masonry homes would take 36 months, almost two years longer than the *Wood Frame* constructions.

This is confirmed by the fact that Wood Frame constructions are partially precast, increasing the quality of the product and minimizing labor.

Figura 9 - Speed of execution in construction methods

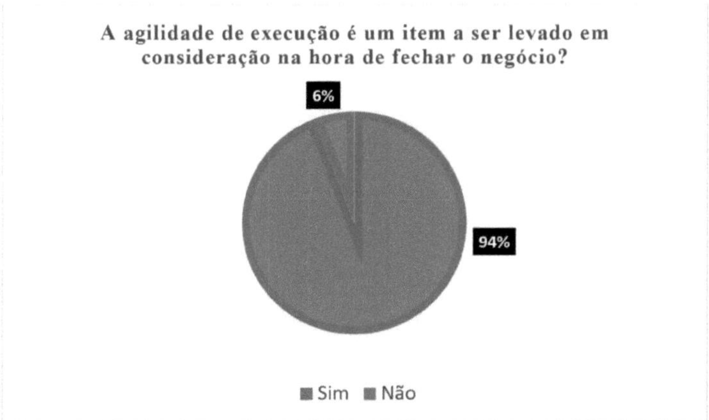

According to Figure 10, 173 people live in conventional masonry houses, showing us once again that the cultural issue makes a difference and really is the main construction option.

In many developed countries such as the United States, Germany and New Zealand, the *Wood Frame* construction method has been consolidated for years. In Brazil, the Conventional Masonry construction system is still the main construction method, one of the main reasons being the embedded culture and the small variety of companies that work with different construction methods.

Figure 10 - Percentage of conventional masonry houses

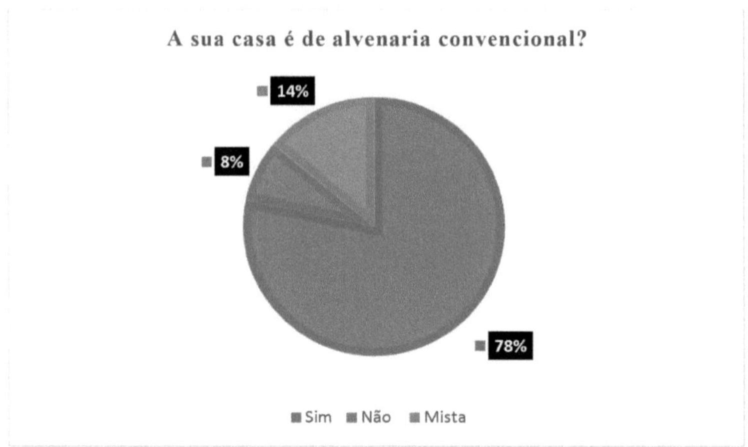

Another benefit is thermal and acoustic insulation. By combining various insulating materials that are poor conductors of heat, such as expanded polystyrene (EPS), glass wool and plasterboard (*drywall*), *Wood Frame* guarantees a significant and noticeable difference in your interiors.

For 61% of the population, conventional masonry houses do not offer good thermal and acoustic insulation (Figure 11).

Figure 7 - Thermal and Acoustic Insulation of Conventional Masonry

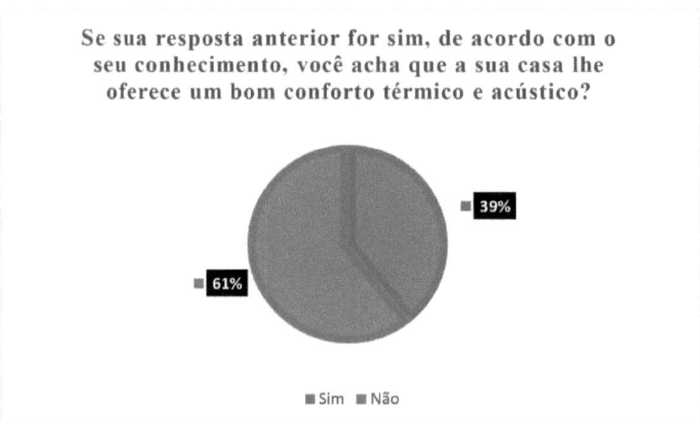

As the region surveyed is very hot in the summer and very cold in the winter, it was important to know which climate was the most unpleasant inside their homes, and

60% of the population perceives that there is a significant difference from outdoors to indoors, both in winter and summer, as shown in Figure 12.

Figura 8 - Internal and external temperature difference

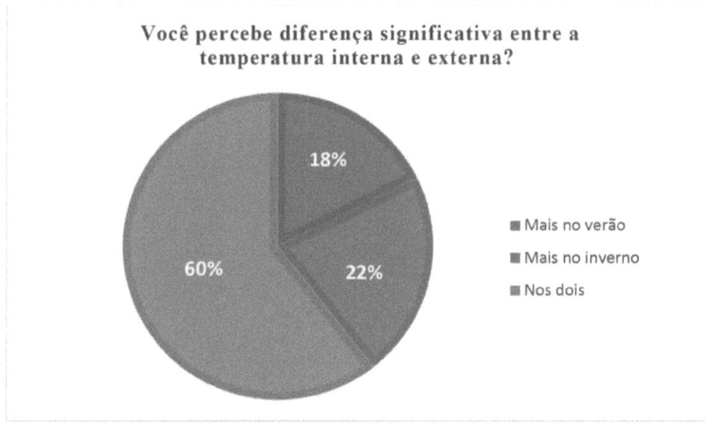

The Brazilian Association of Technical Standards (ABNT) Project 02:135.07-001/2 governs the thermal performance of buildings, covering methods for calculating the thermal transmittance, thermal capacity, thermal lag and solar factor of building elements and components, presents a Table 2 on the thermal properties of materials, comparing the apparent mass density, thermal conductivity and specific heat of materials used in conventional masonry constructions and materials used in *Wood Frame* constructions.

Table 2 - Thermal properties of materials

Material	(")(kg/m^3)	(.=)(W/(m.K))	(c)(kJ/(kg.K)	Application
Mortar				
Ordinary mortar	1800-2100	1,15	1	Masonry
Plaster mortar (or lime and plaster)	1200	0,7	0,84	Masonry
Ceramics				
Clay bricks and tiles	1000 -1300	0,7	0,92	Masonry

	1300-1600	0,9	0,92	
	1600-1800	1	0,92	
	1800-2000	1,05	0,92	
Fiber cement				
Fiber cement boards	1800-2200	0,95	0,84	Masonry
	1400-1800	0,65	0,84	
Concrete (with stone aggregates)				
Normal concrete	2200-2400	1,75	1	Masonry
Plaster				
Designed	1100-1300	0,5	0,84	Masonry and *Wood F.*
Gypsum board or plasterboard	750-1000	0,35	0,84	Masonry and *Wood F.*
Thermal insulation				
Glass slab	10-100	0,045	0,7	*Wood Frame*
Molded expanded polystyrene	15-35	0,04	1,42	*Wood Frame*
Wood and wood products				
Wood	800-1000	0,29	1,34	*Wood F.* e Alvenaria
Oak, ash, pine, cedar, pine	600-750	0,23	1,34	*Wood F.* e

				Alvenaria
	450-600	0,15	1,34	
	300-450	0,12	1,34	
Wood fiberboard (dense)	850-1000	0,2	2,3	*Wood Frame*
Fiberboard wooden (light)	200-250	0,058	2,3	*Wood Frame*
Plywood	450-550	0,15	2,3	*Wood Frame*
	350-450	0,12	2,3	

Source: ABNT, Project 02:135.07-001/2,

*Apparent Mass Density (")(kg/m^3)   **Thermal conductivity (.=)(W/(m.K))

***Specific heat of materials ©(kJ/(kg.K))

Table 2 shows representative data between the *Wood Frame* and Conventional Masonry systems, firstly on apparent mass density (kg/m^3), where the most commonly used materials in masonry constructions are common mortar (1800-2100 kg/m3), bricks (10002000 kg/m3, depending on the type of brick) and normal concrete (2200-2400 kg/m3). In *Wood Frame* constructions, the essential materials for the execution of the system are wooden uprights (300-750 kg/m3, depending on the wood used), *drywall* (750-1000 kg/m3), and the glass slab used in both walls and ceilings, which has an apparent mass density of (10-100 kg/m3).

Evaluating this data, it can be guaranteed that the conventional masonry construction method becomes much heavier, given that all the main materials used in *Wood Frame* constructions do not exceed the kg/m3 of normal concrete, so lightness is one of the great advantages of *Wood Frame* homes.

The proportionality factor k (thermal conductivity), which arises from Fourier's Law, is an individual property of each material and indicates the greater or lesser ease with which a material conducts heat. The numerical values of k vary over a wide range depending on the chemical composition, physical state and temperature of the materials. When the value of k is high, the material is classified as a thermal conductor and otherwise a thermal insulator (SANTOS BAPTISTA, 2013). Thermal conductivity is therefore an important item when choosing materials for construction and other applications, and in many cases it is essential to know how to choose insulating materials, i.e. those with low thermal conductivity, such as polymers.

In conventional masonry constructions, the common mortar used has a thermal conductivity of 1.15 W/(m.K), brick varies from 0.7 to 1.05 W/(m.K) and normal concrete has a thermal conductivity of 1.75 W/(m.K). In *Wood Frame* constructions, the wooden uprights have a thermal conductivity of 0.12 to 0.23 W/(m.K) depending on the wood, the plasterboard 0.35 W/(m.K) and the glass slab 0.045 W/(m.K), these being the materials that make up practically the entire wall, with only the external coatings and the water-repellent membrane missing.

From this, it can be concluded why *Wood Frame* walls are more thermally comfortable than conventional masonry walls: a masonry wall practically has a thermal conductivity of 3.6 W/(m.K) and a *Wood Frame* wall has a thermal conductivity of 0.625 W/(m.K).

Specific heat is defined as the amount of heat given to a substance, resulting in a thermal variation, so the higher the specific heat of the materials, the greater the amount of heat supplied (ALMEIDA, 2012).

In the frame, the common mortar has a specific heat of 1 (kJ/(kg.K)), the brick 0.92 (kJ/(kg.K)), thus adding up to 1.92 (kJ/(kg.K)) of specific heat for the walls. In the *Wood Frame* construction system, the walls are made up of the wooden uprights, which have a specific heat of 1.34 (kJ/(kg.K)), the plasterboard (*drywall*) 0.84 (kJ/(kg.K)), and the glass slab, 0.7 (kJ/(kg.K)), leaving aside the external cladding and the water-repellent membrane, giving a total specific heat of 2.88 (kJ/(kg.K)). Therefore, much more external heat will be needed for the *Wood Frame* walls to begin to vary thermally.

Also evaluating these differences between the methods, the response of people who have lived in *Wood Frame* houses was extremely important. 31% of respondents think that thermal and acoustic comfort is the most attractive feature compared to conventional masonry houses, 15% prefer the speed of project execution, 11% prefer the ease of renovation and maintenance, 8% prefer the internal and external finishes, but the largest number of respondents, i.e. 35% of those interviewed, chose all the answers, agreeing that all these advantages are better than conventional masonry (Figure 13).

Figura 9 - Comparison of construction systems

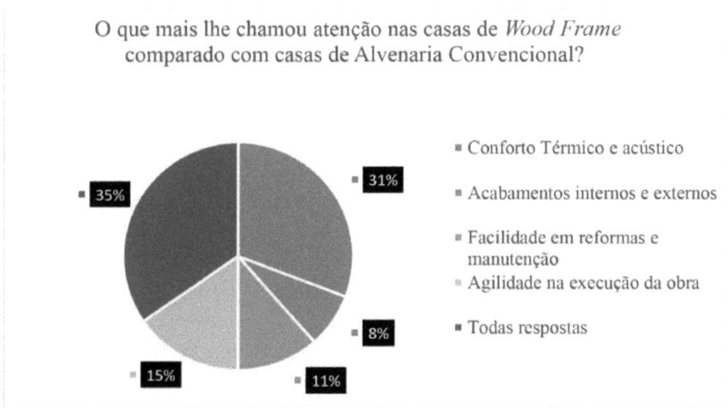

One of the questions asked throughout the work was whether the *Wood Frame* method was really resistant, and according to Vasques and Pizzo (s.d) the structural behavior of *Wood Frame* is superior to that of masonry, in terms of mass, resistance, thermal and acoustic comfort. In *Wood Frame*, moreover, each element receives stresses of different kinds, always in conjunction with other elements, and *Wood Frame* houses present excellent hyperstatic structures.

Most people had no knowledge of *Wood Frame* houses, so they couldn't answer which method would be more resistant to natural disasters (Figure 14).

Figura 10 - Strength of construction systems

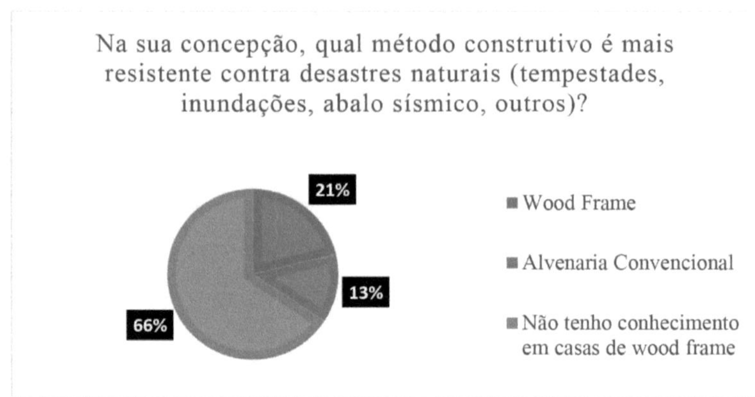

According to 218 respondents (98.64%), sustainability is a subject that is now being taken up by a large number of companies and consumers, and sustainable attitudes are something to be taken into account when choosing a construction method. Industrialized production is based on the manufacture of materials and elements produced in industry, then adapted to the building site. This process is associated with the principles of organization, planning, quality and control, aimed at minimizing waste, increasing productivity and reducing costs (DONIAK, 2014).

Wood is an abundant and renewable material in nature. Trees have the ability to absorb CO_2 from the atmosphere, reducing the greenhouse effect and releasing oxygen. Another characteristic of wood is its low energy production costs compared to other materials (CAMPOS, 2006).

Conventional masonry constructions generate greater energy consumption and polluting gases from the mining of raw materials. *Wood Frame,* on the other hand, reduces CO_2 (carbon dioxide) emissions into the atmosphere by up to 73% (THEODOZIO JÙNIOR, 2006).

For 132 people (59.72%), energy expenditure is a very important issue and must be taken into account in order to adopt sustainable attitudes, saving energy in small areas, such as the use of air conditioning systems, which is essential in the region. Since the thermal conductivity of conventional masonry walls is very high, this number already decreases considerably when it comes to *Wood Frame* constructions, thus minimizing the interior temperature of homes. Another 80 respondents (36.19%) think it is important to review these energy attitudes.

This is the path that the *Wood Frame* construction method follows, forming part of the Sustainable Energy Construction (CES) system, complying with construction standards that are essential for its realization (LEITE; LAHR, 2016).

After explaining the system and all the questions asked, it was also necessary to find out whether the public would choose to build *Wood Frame* houses or would still prefer conventional masonry houses, and after all the knowledge acquired through the information in the explanatory video, 145 people, or 64% of the respondents, would choose the *Wood Frame* construction method (Figure 15).

Figura 11 - Conventional Masonry or Wood Frame

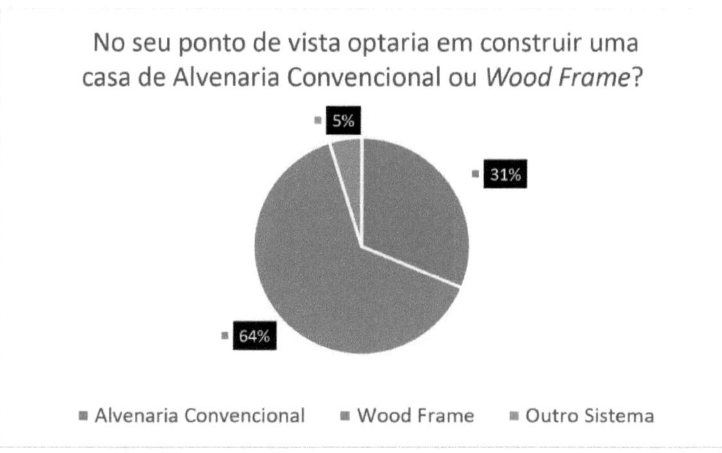

A major impasse in Brazilian construction is the final disposal of the waste that accumulates during the work. Much of this waste ends up being deposited in unsuitable places, creating environmental problems. The manufacture of industrialized materials, such as plasterboard, glass, wood, among others, ensures that waste is minimized and the control and quality of the final system is increased.

4.2 Costing and physical-financial schedule

The aim of drawing up construction budgets is generally to research prices and the inputs used throughout the project, reducing the points of uncertainty in decision-making, analyzing the economic viability of the project and the return on investment (SANTOS; SILVA; OLIVEIRA; 2012).

The *Wood Frame* construction method has numerous advantages over the conventional masonry construction method. One of its main advantages is its total cost combined with productivity, since *Wood Frame* houses are easier and faster to build than conventional masonry houses.

[2]A conventional masonry construction system was verified by analyzing the budget attached as Annex B, which was designed by a construction company in the municipality of Sao Joao do Oeste, totaling R$ 84,998.18, so the cost per square meter is R$ 1,320.26.

Based on the conventional masonry budget, an Appendix A budget was drawn up for the *Wood Frame* construction system, with the same 64.38m2, totaling R$68,949.55, resulting in a cost per square meter of R$1,070.98.

The final price of the *Wood Frame* construction method is 23.28% lower than the final price of the conventional masonry construction method.

One of the significant differences between the stages was in the superstructure, where the *Wood Frame* method evaluates only the wooden uprights, resulting in R$2,401.20, and the same stage in the conventional masonry system showed a result of R$10,429.99, evaluating the reinforced concrete for beams and pillars, the precast slab and the concrete floor. In the *Wood Frame* system, this concrete floor does not need to be made, since it is already made in its infrastructure, the *radier,* which serves as a subfloor to continue the construction.

In the internal and external cladding stages, there was also a significant difference in total costs of R$14,694.36. In the *Wood Frame* budget, wood *sidings* were used as the external cladding, although it is possible to use other types of cladding, such as aluzinc sheets, exposed brick, cement slabs, ceramic slabs, stones and others. For the internal cladding, ceramic tiles have been used in some areas, but it is also possible to use other materials. Sheets of plasterboard (*drywall*) are added to the walls and panels, i.e. to close off the wooden uprights.

Subsequently, in order to demonstrate the execution time for each system, a physical-financial schedule was drawn up (Appendix B), which according to Dias (2004), is the graphical representation of the plan for executing the work and must cover all the phases of execution, from the general preliminary services, through all the activities provided for in the project, to the final cleaning.

The physical and financial schedule for conventional masonry showed that the work would be finished in six months. The *Wood Frame* schedule, based on technical visits, stipulated two months, and the time for a 64.38m house^2 would be a maximum of 20 days, but due to weather conditions, the probable execution time is increasing, as shown in figure 16 below.

Figure 16 - Execution time for construction methods

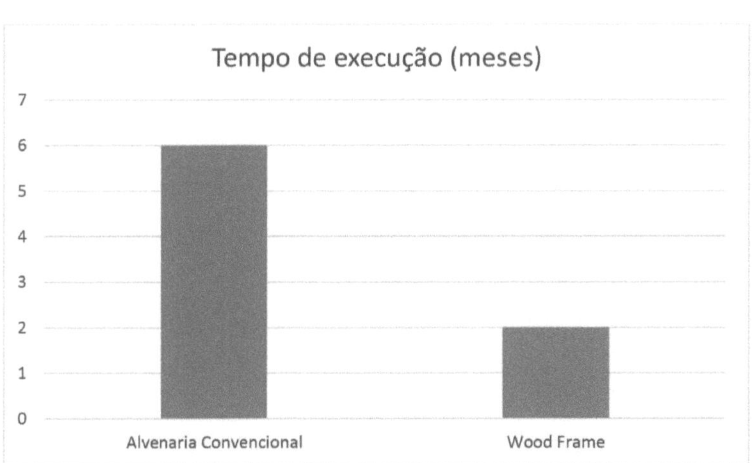

It can be seen that conventional masonry constructions take three times longer to complete than *Wood Frame* constructions, since wood is a very light material compared to concrete and can be easily transported from one place to another, and can even be assembled in sheds, without having to worry about climatic conditions affecting or delaying construction, which is another positive point in the *Wood Frame construction* method.

This would be a solution to meeting the country's housing demand more quickly, since today in Brazil, the Ministry of Planning, Development and Management (PAC), developed by the Federal Government, is seeking infrastructure investments to serve millions of Brazilians, which is fundamental for the country's economy. But the lack of adequate housing is a problem that affects a large part of the Brazilian population, especially the low-income population. One of the causes of this housing deficit is the lack of social housing policies, which has a negative impact on all other areas, such as education, health and security.

Little by little, Brazilian civil construction has seen the emergence of methods capable of combining speed and low cost, with great advantages compared to conventional construction methods. The use of alternative materials or innovative systems can serve as a support for countries like Brazil, solving problems such as the housing deficit.

5 FINAL CONSIDERATIONS

According to the data collected in the questionnaire, it can be concluded that the acceptability of the *Wood Frame* construction system in the Far West region of Santa Catarina is good, with 64% of respondents choosing *Wood Frame*, even though this system still has the structural disadvantages of up to five floors and the need for skilled labor.

However, acceptability meets the needs of customers, as this construction system meets quality, productivity, price and deadline requirements, satisfying the buyer's expectations. Costs are around 23.28% less with the *Wood Frame system*, which in most cases makes all the difference when choosing a construction method, since it already has the same financing power as conventional masonry.

The cost per m^2 of *Wood Frame* today is around R$ 1,070.98 with all the basic items, and that of conventional masonry is around R$ 1,320.26. Remember that changing or eliminating certain materials can make a big difference to the final budget, depending exclusively on the client's needs.

The agility of construction is an extremely important factor, since many people need housing quickly or even to get out of renting and into their own home. These numerous advantages of the *Wood Frame* construction system are combined with the ideas of the *Lean Construction* philosophy, as they guarantee the principles of less human effort (lightweight construction system), less time and less waste (industrialized system, which is consequently faster and generates less waste).

It can therefore be concluded that innovative construction methods such as *Wood Frame* can contribute greatly to the problems affecting Brazil. In order to do so, public policies and company awareness are needed to increase the acceptance of these new construction methods. The importance of preserving, caring for and avoiding waste is clear, so that rates such as environmental impact and homelessness fall, and productivity, sustainability and quality increase.

In this respect, the country of New Zealand is already making great strides, as the system is almost exclusive in the country, with only the use of concrete for the execution of the *radier, as* well as being supported by standards for the system.

6 REFERENCES

ARAÙJO, Luis Otâvio Cocito de. **Civil Construction 1** . Notas de Aula. Rio de Janeiro: UFRJ, 2012.

ASSOCIAÇAO BRASILEIRA DA INDÙSTRIA DE MADEIRA PROCESSED MECHANICS (ABIMCI). Fórum Nacional das Atividades de Base Florestal.

Sectoral study 2013 .

BRAZIL ASSOCIATION OF TECHNICAL STANDARDS (ABNT). **NBR 6122** : foundation design and execution. Rio de Janeiro, 1996. Available at: < http://docente.ifrn.edu.br/valtencirgomes/disciplinas/construcao-de-edificios/nbr-06122-1996-projeto-e-execucao-de-fundacoes >. Accessed: August 18, 2016.

BRAZIL ASSOCIATION OF TECHNICAL STANDARDS (ABNT). **Projeto 02:135.07-001/2:** Desempenho thermalo de edificaçôes Part 2: methods of calculation of thermal transmittance , thermal capacity, thermal retardation and solar factor of building elements and components. Rio de Janeiro, 2003. Available at: < http://www.labeee.ufsc.br/sites/default/files/projetos/normalizacao/Termica_parte2_sE   T2004.pdf >. Accessed: June 1, 2017.

BARROS, Carolina. **Apostila de fundaçôes** : técnicas construtivas, Edificaçôes. Pelotas: IFRS, 2011. Available at:

< https://edificacoes.files.wordpress.com/2011/04/apo-fundac3a7c3b5es- completa.pdf >. Acesso em: 03 de out, 2016.

BAsTOs, Paulo sergio dos santos. **Fundamentos do concreto armado** . Graded by the class. Bauru: UNESP, 2006. Available at:

< http://coral.ufsm.br/decc/ECC1006/Downloads/FUNDAMENTOS.pdf >. Acesso em: 11 de out, 2016.

BASTOS, Paulo Sergio dos Santos. **Foundation shoes** . Graded by the class. Bauru: UNESP, 2016. Available at:

< http://wwwp.feb.unesp.br/pbastos/concreto3/Sapatas.pdf >. Acesso em: 11 de out, 2016.

BERTOLINI, Hibran Osvaldo Lima. **Construçâo via obras secas as factor of productivity and quality.** 2013. 98 f. Trabalho de Conclusao de Curso (Graduaçao em Engenharia Civil)-Departamento de Construçao Civil, Universidade Federal do Rio de Janeiro, Rio de Janeiro, 2013.

BROCK, L. The contemporary brick wall. in: iNTERNATiONAL BRiCK AND BLOCK MASONRY CONFERENCE, 10., 1994, Calgary - Canada. **Proceedings** ... Calgary: iBMAC, 1994, v. 2, pp. 857-865.

JUNIOR CALiL, Carlito; LAHR, Francisco Antonio Rocco; Dias, Antonio Alves. **Dimensioning of madeira structural elements.** 1 ed. Barueri: Editora Manoele, 2003.

CALIL JUNIOR, Carlito; MOLINA, Julio Cesar. **Sistema construtivo em *wood frame* for madeira houses** . Ciências Exatas e Tecnológicas, Londrina, 2010. Available at: < http://www.uel.br/revistas/uel/index.php/semexatas/article/viewFile/4017/6906 >. Acesso em: 22 de out. 2016.

CAMPOS, Rubens Junior Andrade de. **Diretrizes de projeto para produçâo de habitaçôes térreas com estructura tipo platforma e fechamento com plaças cimenticias** . 2006. 165 f. Dissertaçao (Mestrado em Engenharia de Edificaçoes e Saneamento)-Departamento de Construçao Civil, Universidade Estadual de Londrina, Londrina, 2006.

CARDOSO, Larrie Andrey. **Estudo do método construtivo wood framing for the construction of habitats of social interest** . 2015. 79 f. Trabalho de Conclusao de Curso (Graduaçao em Engenharia Civil)-Centro de Tecnologia Engenharia Civil, Universidade Federal de Santa Maria, Santa Maria, 2015.

CARVALHO, Claudio Elias; REIS, Lineu Belico dos; FADIGAS, Eliane A. Amaral.

Energy, resource natural resources practice sustainable development . Barueri: Editora Manoele, 2005, 400 p.

CORREA, Edilson. **Spatial configuration of a reinforced concrete structure to look like conventional masonry** . Available at: < http://engenheiroedilson.wordpress.com/projetos/residencia-4000m ^{2} / >. Access em: 17 out. 2016.

DIAS, Paulo Roberto Vilela. **A budget methodology for civil works** . 9 ed. Rio de Janeiro: Sindicato dos Editores de livros, 2011.

DOCE THE WORK. **Telhas *Shingle*** : o que é, advantages, modelos e preço. [sd]. Available at: < http://casaeconstrucao.org/materiais/telhas-shingle/ >. Acesso em: 04 Nov. 2016.

DONIAK, iria Licia Oliva. **Sistemas construtivos Industrializados** . ABCIC, 2014.

Available at : < http://sindusconsp.com.br/envios/2014/eventos/CONSTRU-BR/Iria%20Doniak.pdf >. Acesso em: 19 de out. 2016.

FERNANDES, André Jorge Ramos Chaves. **Lean construction e construçao sustentâvel:** um estudo de caso. 2015. 88 f. Dissertaçao (Mestrado em Gestao de Qualidade)-Faculdade de Ciência e

Tecnologia, Universidade Fernando Pessoa, Porto, 2015.

FERREIRA, Augusto Sendtko. **Estudo comparativo de sistemas construtivos industrializados:** paredes de concreto, steel frame e wood frame. 2014. 62 f. Trabalho de Conclusao de Curso (Graduaçao em Engenharia Civil)-Centro de Tecnologia curso de Engenharia Civil, Universidade Federal de Santa Maria, Santa Maria, 2014.

FRANKLIN JUNIOR, Ivan; AMARAL, Tatiana Gondim do. **Technological Innovation and Modernization in the Civil Construction Industry** . Encontro Nacional de Engenharia de Produçao, Rio de Janeiro, 2008. 13 p. Available at:

< http://www.uniempre.org.br/user-files/files/enegep2008tnsto86572_10715.pdf >. Acesso em: 01 de out, 2016.

FRANCO, Luiz Sergio. **Desenvolvimento de um construtivo método de alvenaria de vedaçâo de blocos de concreto cellular autoclavados:** proposiçao do método construtivo. Boletim Técnico. Sao Paulo: USP, 1998. 23 p.

FUTURENG. **Wood Framing** . Available at: <http://www.futureng.pt/wood-framing>. Access em: 26 out. 2016.

HYDROFIBER. **Telhas em fibra de vidro** . Araçariguama, [sd]. Available at: < http://www.hidrofiber.com.br/Telha-de-fiberglass.htm >. Acesso em: 04 Nov. 2016.

INSTITUTO BRASILEIRO DE GEOGRAPHIA E ESTATISTICA (IBGE). **Tax of Variation** : setores e construçao civil. Brasilia: Câmara Brasileira da Indùstria da Construçao - CBIC, 2016.

KOSKELA, Lauri. **Application of the New Production Philosophy to Construction** .

Technical Report. Finland: CIFE, 1992. 81 p. Available at: < http://citeseerx.ist.psu.edu/viewdoc/download?doi=10.1.1.15.9598&rep=rep1&type=p df >. Acesso em: 22 de out. 2016.

LEITE, Januària Cecilia Pereira Simoes; LAHR, Francisco Antonio Rocco. **Basic guidelines for project in wood frame** . Available at: < www.fumec.br/revistas/construindo/article/download/4017/1998 >. Acesso em: 28 de set, 2016.

MARTINS, Valdemar; ECKER, Taienne Winni Paiz. **Comparativo dos sistemas construtivos Steel Frame and Wood Frame for habitats of social interest** . 2014. 154 f. Trabalho de Conclusao de Curso (Graduaçao em Engenharia Civil)-departamento acadêmico de Construçao Civil, Universidade Tenológica Federal do Paranà, Pato Branco, 2014.

MEDEIROS, Jonas Silvestre. **Construçao - 101 questions and answers:** project tips, materials and techniques. Barueri: Minha Editora, 2012. 120 p.

MOLITERNO, Antonio. **Caderno de Projetos de telhados em estructuras de madeira** . 4 ed. Sao Paulo: Blucher, 2010. 284 p.

MORAES, Paulo Thiago Araujo; LIMA, Maryangela Geimba. Levantamento e anàlise de procesos construtivos industrializados sob a ótica da sustentabilidade e performance. In: Encontro de Iniciaçao Cientifica e Pós Graduaçao do ITA, 10., Sao José dos Campos, 2009. **Anais..** Sao José dos Campos - SP: ENCITA, 2009, p. 3.

NAKAMURA, Juliana. **PEX tubing** . ed 60. 2013. Available at: < http://equipedeobra.pini.com.br/construcao-reforma/60/tubulacao-pex-conheca-os-principes-componentes-de-instalacoes-hidraulicas-289938-1.aspx >. Acesso em: 04 Nov. 2016.

NET ZERO ENERGY CAPE CODE. **Installation of shingles & siding** . 2011. Available at: < http://netzeroenergycapecod.blogspot.com.br/2011_04_01_archive.html >. Access: 28 out. 2016.

NEW ZEALAND STANDARD. **NZS 3604:1999 Timber Framed Buildings** . 1999.

OLIVEIRA, Gustavo Ventura. **Anâlise comparativa entre o sistema construtivo em light steel framing eo sistema construtivo tradicionalmente emplodo no nordeste do Brasil aplicados a construçâo de casas populares** . 2012. 78 f. Trabalho de Conclusao de Curso (Graduaçao em Engenharia Civil)-Departamento de Engenharia Civil e Ambiental, Universidade Federal da Paraiba, Joao Pessoa, 2012.

PEREIRA, Manuel Fernando Paulo. **Anomalias em paredes de alvenaria sem funçâo structural** . 2005. 489 f. Dissertaçao (Mestrado em Engenharia Civil)-Escola de Engenharia, Universidade do Minho, Guimaraes, 2005.

PFEIIL, Walter; PFEIL, Michele. **Madeira structure** . 6 ed. Rio de Janeiro: LTC Publishers, 2003, 241 p.

PINTO, Jorge Manuel Fonseca. *Lean Construction* : **proposed construction** project evaluation methodology . 2012. 145 f. Dissertaçao (Mestrado integrado em Engenharia Civil)-Departamento de Engenharia Civil, Faculdade de Engenharia da Universidade do Porto, Porto, Portugal, 2012.

PORTO, Thiago Bonjardim; FERNANDES, Danielle Stefane Gualberto. **Basic reinforced concrete course** : according to nbr 6118/2014. Sao Paulo: Oficina de textos, 2015, 51 p.

REDE TECVERDE . **Vantagens da Tecnologia em** *Wood Frame* . Available em: < http://www.redetecverde.com.br/tecnologia >. Access: 28 out. 2016.

DA MADEIRA MAGAZINE. **Civil Construction** : estudo avalia custos de diferentes sistemas de edificaçôes de casas. 137, ed. 2013. Available em: < http://www.remade.com.br/br/revistadamadeira_materia.php?num=1711&subject=Cons

tru%E7%E3o%20Civil&title=Estudo%20avalia%20custos%20de%20diferentes%20sist emas%20de%20edifica%E7% E3o%20de%20casas >. Acesso em: 02 de out. 2016.

IPÊ AMARELO MAGAZINE . **Everything you need to know about** *Wood Frame* : houses made of Madeira that are much more than what you imagine. 2016. Available at: < http://ipeamarelo.cc/tudo-que-voce-precisa-saber-sobre-wood-frame-casas-de- madeira-que-vao-muito-alem-do-que-voce- imagine/ >. Acesso em: 22 de out. 2016.

ROLIM Engenharia Ltda. The *lean* **philosophy** . Fortaleza, 2012. Available: < file:///C:/Users/Ponto%20Fio/Downloads/FILOSOFIA%20LEAN%20(3).pdf > Accessed: 18 de out, 2016.

SACCO, Marcelo de Freitas; STAMATO, Guilherme Corrêa. Light Wood Frame - Construçôes com estructura leve de madeira. **TÉCHNE** : Sao Paulo, 140 ed., Nov. 2008. Available at: < http://techne.pini.com.br/engenharia-civil/140/artigo287602-3.aspx >.

Access em: 28 set. 2016.

SANTOS, Danilo Gonçalves. **Estudo comparativo entre o sistema construtivo traditional e painéis pré moldedados tipo jet casa** . 2014. 39 f Projeto Final (Graduaçao em Engenharia Civil)-Departamento de Engenharia Civil, Universidade Federal de Goiàs, Anâpolis, 2014.

SANTOS, Larissa Carrera Fernandes dos. **Avaliaçao de impactos environmentalas da construçao:** comparaçao entre sistemas construtivos em alvenaria e wood light frame. 2012. 80 f. Monograph (Especializaçao em Construçôes Sustentàveis)-Departamento de Construçao Civil, Universidade Tecnològica Federal do Paranà, Curitiba, 2012.

SANTOS, Ana Paula Santana back; SILVA, Nilmara Delfina yes; OLIVEIRA, Vera Maria de. **Orçamento na construçao civil como instrumento para participaçao em processo licitatório** . 2012. 123 f. Trabalho de Conclusao de Curso (Graduaçao em Ciências Contàbeis)-Centro Universitàrio Católico Salesiano Auxilium, Lins, 2012.

SANTOS BAPTISTA, Fabricio dos. **Lei de Fourier** . 2013. Available at: < https://fenomenosdetransporte2unisul.wordpress.com/2013/06/17/lei-de-fourier-3/ >. Accessed: 01 Jun 2017.

SCHAKELFORD, James F. **Introduçao a ciencia dos materiais para engenheiros** . Sao Paulo: Pearson Prentice Hall, 2008, 575p.

SILVA, Anderson. **Comportamento diaphragma de paredes de madeira no sistema leve plataforma** . 2004. 158 f. Dissertaçao (Mestrado em Engenharia Civil) - departamento Engenharia das Estruturas, Universidade Federal de Uberlândia, Uberlândia, 2004.

SILVA, Carolina Buss yes; SOUZA, Cassiano Donato yes; MORAES, Igor de Paula;

et al. **Sistemas construtivos em *Wood Frame***. October, 2012.

SILVA, Fernando Benigno yes. Constructive systems, *wood frame* - construções com perfis e chapas de madeira. **TÉCHNE** : Curitiba, 161 ed., ago. 2010. Available at: < http://techne.pini.com.br/engenharia-civil/161/sistemas-construtivos-286726-1.aspx >. Access em: 24 out. 2016.

SOUZA, Laurilan Gonçalves. Analyse comparativa do custo de uma unifamiliar casa nos sistemas construtivos de alvenaria, madeira de lei e *woodframe*. **Especialize Magazine:** Florianópolis, 4th ed., Jan. 2013. Available at:

< file:///C:/Users/Ponto%20Fio/Downloads/80c5f1f09008d87d427f2c446ae349e7.pdf >. Access em: 28 set. 2016.

SWELL, Engenharia em Salas Limpas. **Boletim *Shaft***. São José dos Campos, 2013.

Available to:

< http://www.swell.eng.br/catalogos/a4943e1578bac8387bd86aa0e5d7d790.pdf >.

Acesso em: 04 Nov. 2016.

TECH. **Light *wood frame*** : construções com estructura leve de madeira. 140, ed. Nov. 2008. Available in:

< http://www.stamade.com.br/artigos/publ01_revista_techne140.pdf >. Acesso em: 20 de out. 2016.

TECVERDE Construções Eficientes. **Teverde presents a 1st building built in sustainable industrialized technology in Brazil** . Available em: < http://www.tecverde.com.br/2016/08/26/tecverde-apresenta-1o-predio-construido-em- tecnologia-sustentavel-industrializada-do-brasil/ >. Access em: 24 out. 2016.

THEODOZIO JÙNIOR, Stachera. **Avaliaçâo de Emissao de CO2 na construçao civil:** um estudo de caso da habitaçao de interesse social no Paranà. 2006. 176 f. Dissertaçao (Mestrado em Tecnologia)-Programa de Pós Graduaçao, Universidade Tecnològica Federal do Paranà, Curitiba, 2006.

THOMAS, Ercio et al. **Code of Practice n° 01:** Masonry of Vedaçao in ceramic blocks. Sao Paulo: IPT - Instituto de Pesquisas Técnologicas do Estado de Sao Paulo, 2009. 72 p. Available at: < file:///C:/Users/Ponto%20Fio/Downloads/113- Codigo_de_Praticas_n_01%20(2).pdf >. Acesso em: 28 sets. 2016 .

VASQUES, Caio CPC F; PIZZO, Luciana MBF Comparison of conventional, **conventional and**

Wood Frame construction systems in single family homes . [sd]. Unilins, Lins.

VILLAR, Francelene Hermida Rezende. **Alternativas de Sistemas Constructivos para condominiums residentialis horizontalis** : case study. 2005. 151 f Dissertaçao (Mestrado em Construçao Civil)-Centro de Ciências exatas e de tecnologia, Universidade Federal de Sao Carlos, Sao Carlos, 2005.

VIVIAN, André Luiz; PALIARI, José Carlos; NOVAES, Celso Carlos . **Productive advantage of the *light steel framing system*** : it allows smooth construction and constructive rationalization . 2015. Available at: < http://www.aea.com.br/blog/vantagem-produtiva- do-sistema-light-steel-framing-da-construcao-enxuta-a-racionalizacao-construtiva/>. Acesso em: 05 de Nov. 2016.

YAZIGI, Walid. **A Building Technique.** 13 ed. Sao Paulo: Editora Pini, 2013. 856 p.

ZAPARTE, Taiara Aparecida. **Estudo e adequaçao dos principes elementos do modelo canadense de construçao em wood frame para o Brazil** . 2014. 86 f. Trabalho de Conclusao de Curso (Graduaçao em Engenharia Civil)-Departamento de Construçao Civil, Universidade Tecnològica Federal do Paranà, Pato Branco, 2014.

APPENDIX A

Budget spreadsheet and physical and financial timetable *Wood Frame;*

Building System	Wood Frame					
AREA CHART						
Housing	64,38m^2		Total to be built		64,38m2	
VALUES AND COSTS						
Item	Services	Unit	Quantity	Unit Cost	Total Cost	Specification
	GENERAL PRELIMINARY SERVICES					
1.1.1	Technical services, projects, fees, expenses initial, provisional installation, cleaning	vb	1,00	R$ 4.250,00	R$ 4.250,00	Budget, copy, permit, license for electrical and hydraulic installations
	Total				R$ 4.250,00	
	INFRASTRUCTURE					
1.2.1	Land clearing	m^2	600	R$ 0,58	R$ 348,00	Mechanized cleaning
1.2.2	Site location	m^2	64,38	R$ 8,98	R$ 578,13	Plank location
1.2.3	Shallow foundation - radier	vb	1,00	R$ 3.429,00	R$ 3.429,00	Radier (fck=20MPa)
1.2.4	foundation waterproofing	vb	1,00	R$ 425,00	R$ 425,00	asphalt paint, two hands
	Total				R$ 4.780,13	
	SUPERSTRUCTURE					
1.3.1	Wooden mounts	m2	115	R$ 20,88	R$ 2.401,20	comes with nails, angle brackets
	Total				R$ 2.401,20	
	WALLS AND PANELS					
1.4.1	Thermal insulation with glass wool	m2	64,38	R$ 56,91	R$ 3.663,87	lã glass e=2.5cm
1.4.2	plasterboard sheets (drywall)	m2	64,38	R$ 70,00	R$ 4.506,60	nails included
1.4.3	water-repellent membrane	unit	3,00	R$ 586,64	R$ 1.759,92	nails included
1.4.4	expanded polystyrene sheets	m2	64,38	R$ 30,00	R$ 1.931,40	nails included
	Total				R$ 11.861,79	
	WINDOW FRAMES					
1.5.1	Entrance door complete	conj	1,00	R$ 726,94	R$ 726,94	solid wood (80x210x3,5cm)
1.5.2	Full internal doors	conj	4,00	R$ 288,75	R$ 1.155,00	smooth semi hollow wood (80x210x3,5)
1.5.3	Windows	m2	6,12	R$ 359,30	R$ 2.198,92	regional wood shutters
1.5.4	Tippers	m2	0,60	R$ 154,21	R$ 92,53	regional wood shutters
1.5.5	Entrance door - balcony	m2	4,20	R$ 388,79	R$ 1.632,92	solid wood (200x210)
	Total				R$ 5.806,30	
	GLASS AND PLASTICS					
1.6.1	Smooth	m2	10,32	R$ 93,10	R$ 960,79	kitchen, bedroom, laundry area
1.6.2	Fantasy	m2	0,60	R$ 77,85	R$ 46,71	bathroom e=6mm
	Total				R$ 1.007,50	
	COVERS					

1.7.1	Thermal insulation with glass wool	m2	64,38	R$ 56,91	R$ 3.663,87	lă glass e=2.5cm
1.7.2	Roof structure	m2	85,38	R$ 52,20	R$ 4.456,84	wooden structure
1.7.3	osb base kit + shingle tiles + nails	m2	85,38	R$ 72,00	R$ 6.147,36	shingle and blanket tiles
1.7.4	gutters and rufor	m^2	"'4	R$ 15,43	R$ 854,82	galvanized sheet
	Total				**R$ 15.122,88**	
	INTERNAL COATING					
1.8.1	ceramics	m2	57,24	26,81	R$ 1.534,60	kitchen, bathroom and laundry area
	Total				**R$ 1.534,60**	
	EXTERNAL COATING					
1.9.1	Wooden siding	unit	51,90	R$ 98,00	R$ 5.086,20	nails included
	Total				**R$ 5.086,20**	
	LINING					
2.1.1	drywall	m2	64,38	R$ 70,00	R$ 4.506,60	nails included
	Total				**R$ 4.506,60**	

VALUES AND COSTS

Item	Services	Unit	Quantity	Unit Cost	Total Cost	Specification
	PAINTING					
2.2.1	Painting the inside of the plaster with acrylic paint	m2	179,01	R$ 11,86	R$ 2.123,06	acrylic sealer and 2x paint
2.2.2	Painting on wood with enamel paint	m2	38,64	R$ 21,63	R$ 835,78	1x matte and 2x paint
	Total				**R$ 2.958,84**	
	FLOORS					
2.3.1	ceramics	m2	59,71	R$ 26,81	R$ 1.600,83	stoneware plaque 45x45cm
	Total				**R$ 1.600,83**	
	ELECTRICAL INSTALLATIONS AND TELEPHONE					
2.4.1	Pipes and boxes in the ceiling	vb	1,00	R$ 254,68	R$ 254,68	according to project
2.4.2	Pipes and boxes in panels	vb	1,00	R$ 271,00	R$ 271,00	according to project
2.4.3	Distribution board	unit	1,00	R$ 310,12	R$ 310,12	according to project
2.4.4	Sockets, switches and circuit breakers	vb	1,00	R$ 234,45	R$ 234,45	according to project
2.4.5	Power input panel	unit	1,00	R$ 180,00	R$ 180,00	according to project
2.4.6	Lamps	unit	12,00	R$ 12,20	R$ 146,40	according to project
2.4.7	threading	vb	1,00	R$ 496,00	R$ 496,00	according to project
	Total				**R$ 1.892,65**	
	HYDRAULIC INSTALLATIONS					
2.5.1	Trestle and water meter	vb	1,00	R$ 320,00	R$ 320,00	according to project
2.5.2	cold water pipes	vb	1,00	R$ 640,00	R$ 640,00	according to project
2.5.3	cold water tank	unit	1,00	R$ 235,00	R$ 235,00	polyethylene or vibro fiber = 500l
	Total				**R$ 1.195,00**	
	SEWAGE AND RAINWATER INSTALLATION					
2.6.1	piping	vb	1,00	R$ 712,00	R$ 712,00	according to project
2.6.2	boxes	unit	3,00	R$ 86,00	R$ 258,00	according to project
2.6.3	septic tank	unit	1,00	R$ 1.242,15	R$ 1.242,15	according to project
2.6.4	sink	unit	1,00	R$ 1.098,00	R$ 1.098,00	according to project

	Total					R$ 3.310,15	
	TABLEWARE AND METALS						
2.7.1	Toilet bowls		unit	1,00	R$ 368,65	R$ 368,65	sanitary ware basin
2.7.2	Washbasins		unit	1,00	R$ 488,57	R$ 488,57	countertop washbasin
2.7.3	Tank		unit	1,00	R$ 126,70	R$ 126,70	fiber tank
2.7.4	Taps and registers		unit	4,00	R$ 40,44	R$ 161,76	3 taps and 1 pressure valve
2.7.5	drawer damper		unit	4,00	R$ 108,25	R$ 433,00	used in water descents
2.7.6	shower		unit	1,00	R$ 56,20	R$ 56,20	plastic shower head
	Total					R$ 1.634,88	
	COMPLEMENTS						
2.8.1	Final cleaning		vb	0,00	R$-	R$-	
	Total					R$ 68.949,55	
	Cost/m²					R$ 1.070,98	

APPENDIX B

Mapping - Comparisons between Conventional Masonry in Brazil and Wood Frame in New Zealand

MAPPING - CONVENTIONAL MASONRY IN BRAZIL VS. WOOD FRAME IN NEW ZEALAND			
Masonry context	**Wood Frame Context**	**Current Conditions Masonry**	**Current Conditions Wood Frame**
Conventional Masonry construction method, the most common building system in Brazil today	Wood Frame construction method, the most common construction system used in New Zealand and many developed countries.	The conventional masonry construction system is a method that generates a lot of waste, which is a major problem worldwide. Another condition is the delay in construction. The excessive use of water throughout the process is also very high.	Innovative construction method, where the total waste does not fill more than one bucket, the use of water is only in the foundation, many inspections are made so that the material used always complies with the adopted standards, and the main thing is that wood is a renewable and abundant material.
Advantages Masonry	**Advantages of Wood Frame**		
Structural capacity Availability of labor Durability Construction standards	Structural lightness Speed of execution Higher productivity Lower waste generation Thermal and acoustic comfort Ease of electrical installation Ease of hydraulic installation Lower energy consumption Lower water consumption Construction standards	**Proposed countermeasures Masonry**	**Proposed Wood Frame countermeasures**
		The use of alternative materials or different methods to replace the masonry method can help countries like Brazil solve problems such as housing shortages, waste generation, excessive water and energy consumption.	Because the system is almost exclusive, there are no countermeasures proposed to the system.
Disadvantages Masonry	**Disadvantages of Wood Frame**	**Conventional Masonry Plan**	**Wood Frame plan**
Specific weight Speed of execution Lower productivity High waste generation Thermal and acoustic comfort	Skilled labor Low structural capacity	Cultural changes; Implementing new construction methods; Demonstrating the advantages of innovative	Because the system is well regulated and executed strictly according to the proposed standards, there is no plan for improvement.

| | | methods; | |
| Electrical installations Hydraulic installations High energy consumption High water consumption | | Raising awareness of sustainability Acceptance by the population Minimizing the housing shortage Creating public policies and stimulating the use of Wood Frame in Brazil | |

Mapping - Comparisons between Wood Frame Brazil and Wood Frame New Zealand;

MAPPING A3 - WOOD FRAME IN BRAZIL X WOOD FRAME IN NEW ZEALAND			
Context Wood Frame Brazil	**Context Wood Frame NZ**	**Current Conditions Wood Frame Brazil**	**Current Conditions Wood Frame NZ**
Wood Frame construction method, a construction system that is not very widespread in Brazil today	Wood Frame construction method, the most common construction system used in New Zealand and many developed countries.	The Wood Frame system is a method that is rarely used in construction. In Brazil, the main cause is the culture of the population, where people do not believe that a home in Wood Frames can be as strong as masonry constructions. It is also a branch of few companies in Brazil.	New Zealand has adopted the Wood Frame construction method as its main building system. It has hundreds of regulations, accompanied by technical inspections by building inspectors, from the compaction of the soil to the final handover.
Advantages of Wood Frame Brasil	**Advantages of Wood Frame NZ**		
Structural lightness Speed of execution Greater productivity Less waste generation Thermal and acoustic comfort Ease of electrical installation Ease of hydraulic installation Energy efficiency Lower water consumption	Structural lightness Speed of execution Greater productivity Less waste generation Thermal and acoustic comfort Ease of electrical installation Ease of hydraulic installation Energy efficiency Lower water consumption Construction standards	**Wood Frame Brazil Countermeasures** The Wood Frame construction system in Brazil is ineffective in terms of standards and regulations; In Brazil, OSB panels are still used to close off the walls. This is an unnecessary item and could be replaced with EPS sheets, i.e. expanded polystyrene.	**Wood Frame NZ Countermeasures** No countermeasures are needed because the system is very consolidated
Disadvantages Wood F. Bra.	**Disadvantages of Wood Frame NZ**	**Wood Frame Masonry Plan Brazil**	**Wood Frame NZ plan**
Insufficient standards Low structural capacity Skilled workforce OSB sheets that don't guarantee as much thermal and acoustic comfort as the expanded polystyrene (EPS) used in New Zealand	Low structural capacity Skilled workforce	Study new products Add standards to this system; Think about adjusting the Brazilian system in line with the system used abroad; Setting up homes under 100m^2 in sheds, knowing that the system is light and easy to transport, with no problems regarding the weather;	Because the system is well regulated and executed strictly according to the proposed standards, there is no plan for improvement.
Observations on Brazilian Wood Frame		**Curiosities Wood Frame Exterior**	
The NBR 15575 performance standard has already been approved for this type of system, in terms of fire, thermal and acoustic performance, among others. August 24, 2016 saw the construction of the first three-storey building, built in 64 hours in Wood Frame (TECVERDE, 2016).		In Japan, an eight-storey Wood Frame building was built that withstood the simulation of a 7.5 magnitude earthquake. (BRASIL ENGENHARIA) In April 2017, New Zealand experienced Cyclone Cook with winds of approximately 160km/h, large trees were uprooted, and the Wood Frame houses remained intact.	

APPENDIX C

Wood Frame Building System in New Zealand

I was on a technical visit to New Zealand for a month in search of knowledge about *Wood Frame* construction. New Zealand is a country that adopts this system in practically all its buildings, making it an excellent reference.

The first stage of the project is the *radier-type* foundation, which is laid with wooden planks or even the concrete block itself, as shown in figure 1. Many people opt for the concrete block because it allows for easy future finishing. It's important to mention that before starting the construction site, the soil is compacted. Once the compaction has been completed, site inspectors go to the site to measure the degree of compaction, which must comply with the stipulated standards.

NZS 4402: Soil test methods for civil engineering purposes;

Figure 12 - Site location with concrete block

After locating the building site, all the necessary piping for hydraulic and electrical installations is planned, as shown in figure 2.

Figure 13 - Hydraulic and electrical installations

Continuing the project, a gravel ballast is placed along the site, allowing for better leveling of the ground. Plastic sheeting is inserted on top of the gravel ballast, which helps waterproofing and prevents the concrete from spreading over the gravel. After applying the tarpaulin, steel mesh is inserted to improve its resistance, as shown in figure 3.

Figure 14 - Application of plastic sheeting and steel mesh

Finally, according to figure 4, the concrete is poured directly on site with a truck mixer. The concrete is then leveled and smoothed with a floor polisher, since in this type of foundation, the *radier* already serves as the subfloor.

Figure 15 - Concreting the Radier

After the concrete has cured, the wooden uprights are taken to the building site, wall by wall according to the project (figure 5).

Figure 16 - Wooden structure

Following on from this, the stage of lifting the timber structures begins, as shown in figure 6, with a workforce of two. The timber is fixed to the *radier* with anchor bolts, practically every two wooden uprights. The spacing of the uprights is between 40 and 60 cm, and they are subjected to structural stress near windows, doors and sides.

Figure 17 - Wood Frame: a) Wood Frame skeleton; b) anchor bolts;

(a) (b)

The wood used in the building structures is *pine* and is autoclaved, guaranteeing protection against termites, rot and other biological decay agents. But in order to improve its performance and lifespan, the wood used in the structures must not come into contact with water under any circumstances, in accordance with its standards and specifications (figure 7):

NZS 3640: 2003 Chemical preservation of wood;

NZS 3602: 2003 Wood-based products and timber for use in construction;

NZS 3603: 1993 Standard timber structures;

NZS 3601: 1973 Dimensions for timber;

NZS 4210: 1989 Code of practice for timber construction: materials and workmanship;

Figure 18 - Types of wood: a) treated plywood; b) wood only used indoors; c) wood treated for water;

(a) (b) (c)

When the walls are finished, the waterproofing membrane is laid, acting as a barrier to water, heat, wind and humidity. Next, the roof structure is screwed and nailed to the wooden guides, finishing with the roof as shown in figure 8, which can be made of various materials, the most common being aluzinc tiles.

NZS 4408:1988 Specification for asphalt shingles made from mineral-grained glass felt;

Figure 19 - Construction stage: a) Water repellent membrane and roof structure; b) roof;

(a)

(b)

The electrical and water installations, as shown in figure 9, are installed between the uprights and wooden guides. The pipes used are cross-linked polyethylene (PEX), as these are flexible pipes that make it easier to make bends and reduce the use of fittings.

Figure 20 - Electrical and hydraulic installation

Between the guides and the uprights of the wall and ceiling, the glass slab is placed, which helps with thermal and acoustic insulation. After that, all the *drywall* sheets are installed inside the house, as shown in figure 10, which have various characteristics such as: good thermal and acoustic insulation, excellent aesthetic results, a lightweight material that is easy to handle, among others.

Figure 21 - Plasterboard (Drywall)

In many buildings, to increase thermal and acoustic insulation, a layer of expanded polystyrene (EPS) is placed on top of the waterproof membrane, as shown in figure 11. In addition to its thermo-acoustic characteristics, it has advantages such as lightness, low water absorption, easy handling, among others. All these stages, from the wooden structure to the EPS closure, are part of the formation of *Wood Frame* walls.

Figure 22 - Expanded Polystyrene sheets: a) EPS-resistant house; b) enlarged EPS figure

(a)   (b)

The external cladding can be made of cement slabs, stone, aluzinc, ceramics, etc., but the most common are wooden *sidings* (figure 12). They have a male-female fitting, so that if water gets in between the cladding and any closing panel, it has a place to escape.

NZS 3611:1970 Specification for exterior cladding;

Figure 23 - External cladding: a) External wall clad with wood siding and cementitious board; b) demonstration of tongue-and-groove wood siding;

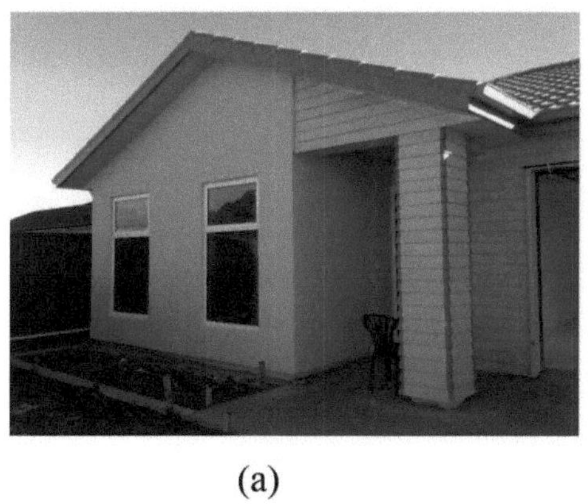 

(a)   (b)

And the last stage of the process is painting, both external and internal (figure 13).

NZS 7703:1985 The painting of buildings;

Figure 24 - Multifamily dwelling with external painting

More
Books!

info@omniscriptum.com
www.omniscriptum.com
OMNIScriptum

Printed by Books on Demand GmbH, Norderstedt / Germany